FUN FACT FILE: INVENTIONS THAT CHANGED THE WORLD

20 FUN FACTS ABOUT THE INTERNET

BY KRISTEN RAJCZAK NELSON

Gareth Stevens PUBLISHING

Please visit our website, www.garethstevens.com. For a free color catalog of all our high-quality books, call toll free 1-800-542-2595 or fax 1-877-542-2596.

Library of Congress Cataloging-in-Publication Data
Names: Rajczak Nelson, Kristen, author.
Title: 20 fun facts about the internet / Kristen Rajczak Nelson.
Other titles: Twenty fun facts about the internet
Description: Buffalo, New York : Gareth Stevens Publishing, [2024] | Series: Fun fact file: inventions that changed the world | Includes index.
Identifiers: LCCN 2022052154 (print) | LCCN 2022052155 (ebook) | ISBN 9781538287651 (library binding) | ISBN 9781538287644 (paperback) | ISBN 9781538287668 (ebook)
Subjects: LCSH: Internet–Juvenile literature.
Classification: LCC TK5105.875.I57 R334 2023 (print) | LCC TK5105.875.I57 (ebook) | DDC 004.67/8–dc23/eng/20221107
LC record available at https://lccn.loc.gov/2022052154
LC ebook record available at https://lccn.loc.gov/2022052155

Published in 2024 by
Gareth Stevens Publishing
2544 Clinton Street
Buffalo, NY 14224

Designer: Leslie Taylor
Editor: Therese Shea

Photo credits: Cover, p. 1 (main) metamorworks/Shutterstock.com; Series Art (file folder) David Smart/Shutterstock.com, (binder clip) luckyraccoon/Shutterstock.com, (wood grain background) ARENA Creative/Shutterstock.com; p. 5 U.S. Census Bureau employees/https://commons.wikimedia.org/wiki/File:Univac_I_Census_dedication.jpg; p. 6 Touch Of Light/https://commons.wikimedia.org/wiki/File:The_Pentagon,_cropped_square.png; p. 7 (Kleinrock) Fred Prouser/Alamy.com, (computer icons) Domofon/Shutterstock.com; p. 8 Fred Prouser/Alamy.com; p. 9 (map) Peter Hermes Furian/Shutterstock.com; p. 10 Semaforo GMS/https://commons.wikimedia.org/wiki/File:Arpanet_in_the_1970s.png; p. 11 (@ symbol image) Monster Ztudio/Shutterstock.com, (Tomlinson) Andreu Veà, WiWiW.org/https://commons.wikimedia.org/wiki/File:With_Ray_Tomlinson.jpg; p. 12 nuttapon averuttaman/Shutterstock.com; p. 13 BrightSpace/Shutterstock.com; p. 14 Red Tiger/Shutterstock.com; p. 15 (Berners-Lee) John S. and James L. Knight Foundation/https://commons.wikimedia.org/wiki/File:Tim_Berners-Lee-Knight-crop.jpg, (mosaic) Bartholomeus Thoth/https://commons.wikimedia.org/wiki/File:Mosaic_at_Sheen_Lane_Centre_honouring_Tim_Berners-Lee_(2017).JPG; p. 16 (Andreessen) JD Lasica/https://commons.wikimedia.org/wiki/File:Marc_Andreessen-9.jpg, (screenshot) Koshiro K/Shutterstock.com; p. 17 CapturePB/Shutterstock.com; p. 18 antoniodiaz/Shutterstock.com; p. 19 carballo/Shutterstock.com; p. 20 PK Studio/Shutterstock.com; p. 21 (Zuckerberg) catwalker/Shutterstock.com, (storefront) Tommyl33/Shutterstock.com; p. 22 wavebreakmedia/Shutterstock.com; p. 23 Miss Ty/Shutterstock.com; p. 24 Krakenimages.com/Shutterstock.com; p. 25 McLittle Stock/Shutterstock.com; p. 26 Giulio Benzin/Shutterstock.com; p. 27 The Bold Bureau/Shutterstock.com; p. 29 NEERAZ CHATURVEDI/Shutterstock.com.

Printed in the United States of America

CPSIA compliance information: Batch #CSGS24: For further information contact Gareth Stevens at 1-800-542-2595.

CONTENTS

Words in the glossary appear in **bold** type the first time they are used in the text.

A CONNECTED WORLD

The internet is a **network** that connects computers worldwide. The earliest computer network connected only a small group of colleges. In 2022, the **United Nations** reported almost 5 billion people have used the internet!

The beginnings of the internet date back to the 1960s. At that point, computers were huge. One computer could fill an entire room! It took many years, but the internet of today came from the work done with these computers.

UNIVAC 1 was a U.S. government computer in the 1950s. It weighed over 16,000 pounds (7,257 kg)!

EARLY DAYS

FUN FACT: 1

THE IDEA FOR THE INTERNET STARTED AT THE U.S. DEPARTMENT OF DEFENSE (DOD).

the Pentagon, DOD headquarters

In 1958, President Dwight Eisenhower approved the Advanced Research Projects Agency (ARPA) at the DOD. ARPA studied how **technology** could help the military. ARPA scientist J. C. R. Licklider wrote about a computer network connecting the world.

The DOD is charged with protecting, or guarding, the nation. The government wanted a way to communicate that could survive an attack. This plan became the internet.

Kleinrock was trying to send a message from a computer at UCLA (University of California, Los Angeles) to a computer at Stanford University in Stanford, California.

Leonard Kleinrock

FUN FACT: 2

THE FIRST MESSAGE SENT BETWEEN COMPUTERS CRASHED THE SYSTEM.

In 1965, Leonard Kleinrock sent the first message between two computers—almost. He tried to send the word "login." The other computer crashed after just two letters were sent!

Computer scientists Lawrence Roberts (left) and Leonard Kleinrock (right) are just two of the many people who had a part in the internet's beginnings.

FUN FACT: 3

ARPANET MADE THE INTERNET POSSIBLE.

ARPA's computer network, or ARPANET, was first used in 1969. Chief scientist Lawrence Roberts had the idea to use "packet switching" in its design, or plan. Packet switching is a way for computers to separate and send **data**.

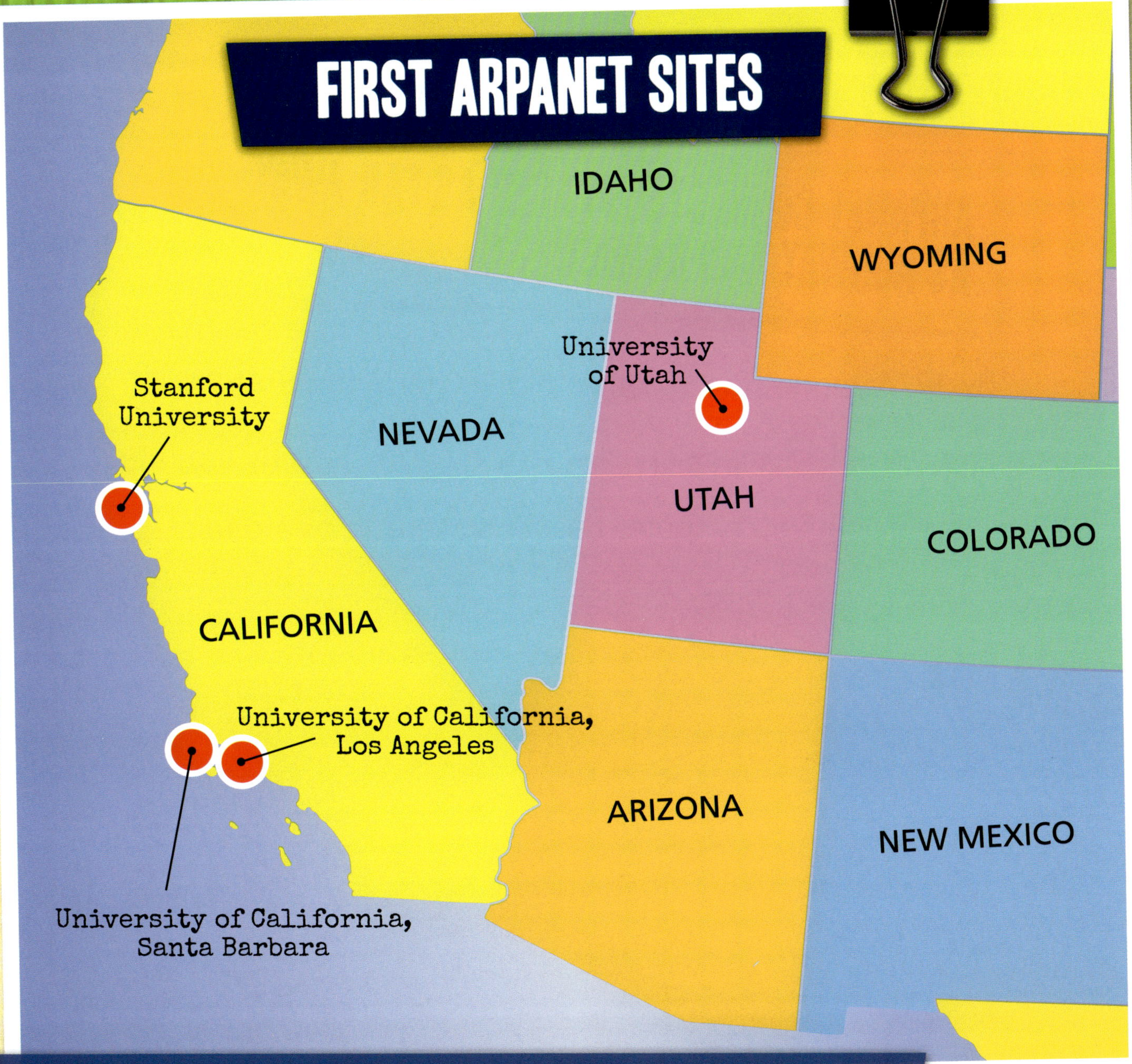

The first four locations connected to ARPANET were in California and Utah.

FUN FACT: 4

ARPANET CONNECTED PLACES AS FAR AWAY AS NORWAY AND HAWAII!

By 1973, more than 30 other places had joined ARPANET. However, ARPANET was still used only by scientists. By the late 1970s, several other computer networks were in operation.

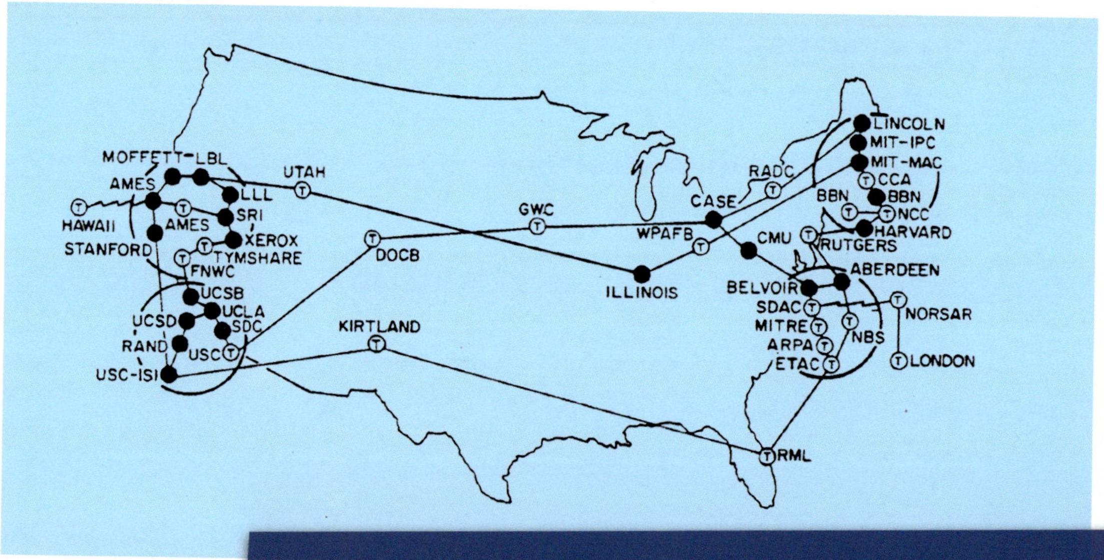

This map shows the locations connected by ARPANET in the 1970s. ARPANET existed until 1990.

Ray Tomlinson also came up with the idea to use the @ symbol, or character, in email addresses.

Ray Tomlinson

FUN FACT: 5

EMAIL IS MORE THAN 50 YEARS OLD!

ARPANET users needed to send messages through the network. Ray Tomlinson created a messaging **program**. He sent the first electronic mail, or email, in 1971. Five years later, British queen Elizabeth II used ARPANET to send the first royal email!

THE INTERNET IS BORN

FUN FACT: 6

JANUARY 1, 1983, IS THE INTERNET'S BIRTHDAY!

At first, computers on different networks didn't have a standard way to communicate. It was like they were talking different languages. Networks changed to a new protocol, or set of rules, on this day. They began to move data in the same way.

The move to all networks using the same protocol was planned for years before it happened.

FUN FACT: 7

WITHOUT DOMAIN NAMES, WEBSITES WOULD BE FOUND USING A LONG STRING OF NUMBERS!

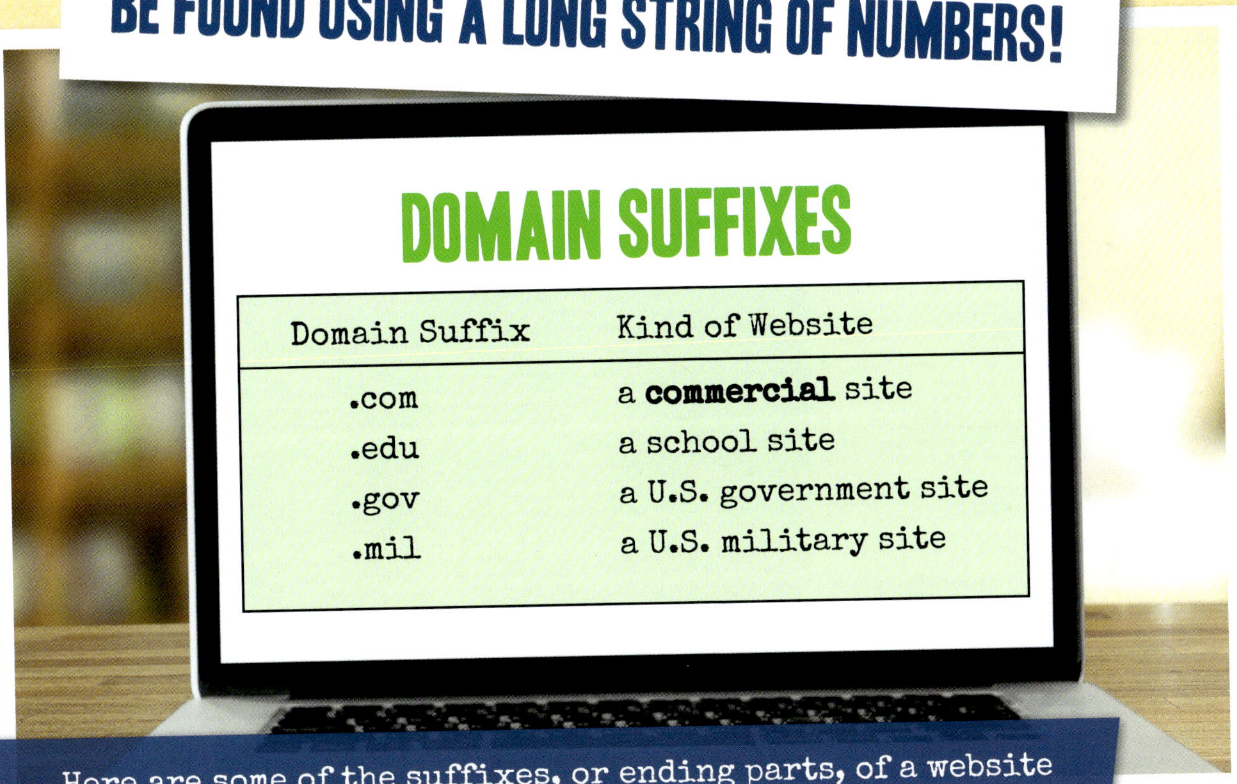

DOMAIN SUFFIXES

Domain Suffix	Kind of Website
.com	a **commercial** site
.edu	a school site
.gov	a U.S. government site
.mil	a U.S. military site

Here are some of the suffixes, or ending parts, of a website domain name. They tell what kind of site you're visiting.

A domain name is the name of a website. It's connected to an IP address, which is a list of numbers that tells which computer is which in a network.

INTO THE WEB

FUN FACT: 8

THE WORLD WIDE WEB IS DIFFERENT FROM THE INTERNET.

The internet is the network that connects computers all over the world. The World Wide Web works on the internet. It's the name for all the pages and content you see online.

Think of the internet as the roads that connect cities. The World Wide Web is the stores and houses built on the roads.

While in Switzerland, Tim Berners-Lee was working on a way for scientists to share data more easily. This was the World Wide Web. He invented other computer technology too.

Tim Berners-Lee

FUN FACT: 9

THE WORLD WIDE WEB WAS BORN IN THE COUNTRY OF SWITZERLAND.

Computer scientist Tim Berners-Lee came up with the idea of the World Wide Web in 1989. But his web could be used only on certain computers. Nicola Pellow, a student, made it work on any computer.

FUN FACT: 10

MORE THAN 99,000 WEBSITES WERE CREATED BETWEEN 1993 AND 1996.

In 1993, the first widely used web browser, Mosaic, was **launched**. A browser is a program to find and look at internet content. Many people began to create their own websites at this time.

Marc Andreessen was a student when he created Mosaic with another programmer. Browsers today include Firefox, Safari, Edge, and Chrome.

An operating system (OS), such as Windows 95, is the **software** in charge of the programs on a computer and how it looks on the screen.

FUN FACT: 11

AMAZON AND EBAY LAUNCHED AROUND THE SAME TIME.

These now-huge businesses went online in 1995. The Microsoft company's Windows 95 also came out that year. It was the first operating system created to be easy to use for everyone.

BOOMING!

FUN FACT: 12

NEW WORDS AND SAYINGS HAVE COME INTO USE BECAUSE OF THE INTERNET.

Some online gamers live stream their games. This means they're sharing the video of their games over the internet as they're playing them.

In 1984, the term "cyberspace" was first used. It means the "online world." By 1992, "surfing the internet" had become a popular saying. The *Oxford English Dictionary* added the term "live stream" in 2021.

People hoping to become famous create online content to "go viral"!

FUN FACT: 13

AN ANIMATED DANCING BABY WENT VIRAL IN 1996.

This was the first viral video! However, the saying "going viral" wasn't really used until 2009. "Going viral" means a video, photo, or story gets shared millions of times quickly all over the world.

FUN FACT: 14

GOOGLE BECAME THE WORLD'S LARGEST SEARCH ENGINE IN JUST TWO YEARS.

The search engine Google launched in 1998. By 2000, it was the most popular. People began to use the company name as a verb, or action word. They "googled" what they wanted to know!

A search engine is a program that matches a user's search words to websites that use those words.

Mark Zuckerberg was one of the founders of Facebook. His website helped make social media popular. Social media are websites that allow people to create online communities to share ideas.

FUN FACT: 15

FACEBOOK STARTED AS A WEBSITE FOR COLLEGE STUDENTS.

When Facebook launched in 2004, only Harvard University students could use it. But soon after, users just needed a college email address to sign up. By 2006, anyone at least 13 years old with an email address could join.

THE INTERNET OF TODAY

FUN FACT: 16

MANY PEOPLE THINK THE INTERNET IS AS IMPORTANT AS ELECTRICITY.

This is why many want "net neutrality." Net neutrality is the idea that companies that provide us internet service should treat all websites the same. They shouldn't allow some websites to work better than others.

Having net neutrality laws would make it illegal for an internet company to slow down internet speeds for some sites.

SOCIAL MEDIA TIMELINE

2003 MySpace, the first worldwide social network, launches.

2004 Facebook launches at Harvard University.

2005 YouTube launches. Reddit is founded.

2006 The first tweet is sent on Twitter.

2009 WhatsApp launches.

2010 Pinterest and Instagram are founded.

2011 The first snaps are sent on Snapchat.

2016 TikTok launches.

New social media sites come out all the time. This timeline shows the launches of some of the biggest today.

FUN FACT: 17

THE FIRST AMENDMENT OF THE U.S. CONSTITUTION PROTECTS MUCH OF THE CONTENT ON THE INTERNET.

The First Amendment includes the right to free speech. In 1996, the U.S. Supreme Court, the highest court in the country, ruled the First Amendment protects content on the internet too.

Free speech doesn't include **threatening** others, trying to get someone to carry out a crime, or saying harmful, untrue things.

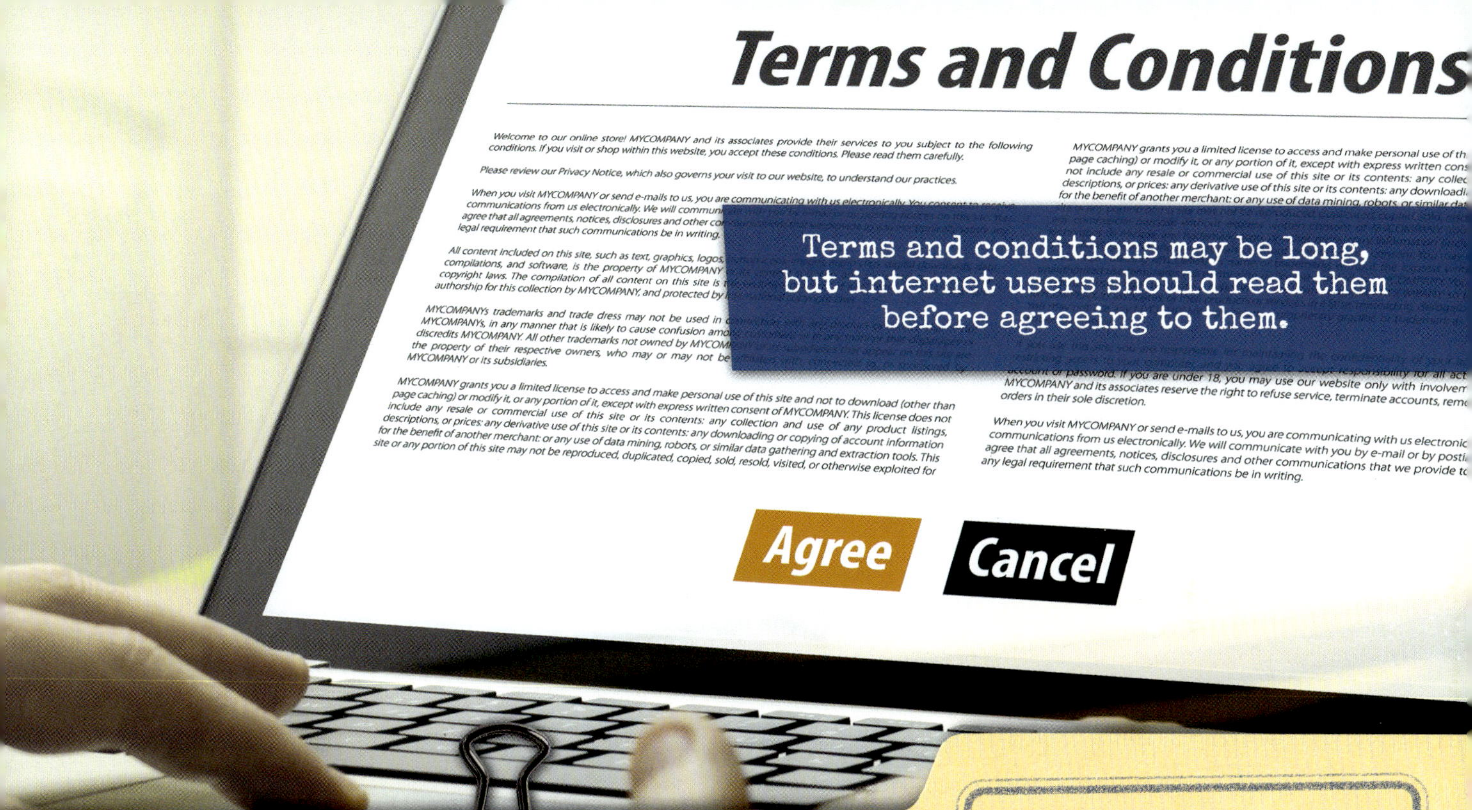

Terms and conditions may be long, but internet users should read them before agreeing to them.

FUN FACT: 18

THINGS POSTED ON THE INTERNET STAY ON THE INTERNET.

Many internet companies, such as Google, Facebook, and YouTube, ask users to agree to certain conditions. These may let them store users' data, including pictures, even if the user deletes, or removes, them.

FUN FACT: 19

TOOTHBRUSHES, REFRIGERATORS, AND CARS CAN CONNECT TO THE INTERNET.

Everyday items connected to the internet are part of the network called the Internet of Things. These items have computer abilities. They record data and are able to reply to the data.

Things that have computing abilities are called "smart."

Satellite internet is a good choice in some areas. People may not live near cables that carry internet to people's homes.

FUN FACT: 20

SOME PEOPLE GET THEIR INTERNET FROM SPACE.

To use satellite internet, a satellite dish on top of a building or house is needed. Data is sent and received using a computer, the dish, and a satellite in space!

A HUMAN RIGHT

Today, nearly 3 billion people on Earth still don't have internet **access**. Access may cost too much or be unavailable where they live. In the United States, more than 27 million homes don't have internet access.

Many people argue that internet access is a basic human right. Access allows people to take part in government. It helps them find out about healthcare and other services. It helps them learn too. Some groups are working hard to spread internet access around the globe.

Scientists are trying to think of ways to bring the internet to more people around the world. Successful ideas have included balloons and beams of light.

GLOSSARY

access: A way of being able to get or use a resource.

amendment: A change or addition to a constitution.

animated: Consisting of drawings or computer images that appear to move.

commercial: Having to do with the buying and selling of goods and services.

data: Information created and stored by a computer.

launch: To begin a computer program.

network: A system using multiple computers or other devices that communicate with each other.

program: A set of computer instructions, or commands.

satellite: A machine sent into space that moves around Earth to complete scientific research or for a type of communication.

software: A set of programs that instructs a computer what to do.

technology: Using science, engineering, and other industries to invent useful tools or to solve problems. Also a machine, piece of equipment, or method created by technology.

threaten: To say you will harm someone or do something unwanted to get what you want.

United Nations: A group of nations that united after World War II with the intention of resolving conflicts between nations.

FOR MORE INFORMATION

BOOKS

Bell, Samantha S. *Before the Internet.* Lake Elmo, MN: Focus Readers/Beacon, 2020.

González, Echo Elise. *The Internet.* Chicago, IL: World Book, Inc., 2020.

Ignotofsky, Rachel. *The History of the Computer: People, Inventions, and Technology That Changed Our World.* Berkeley, CA: Ten Speed Press, 2022.

WEBSITES

Tim Berners-Lee and the Invention of the Internet for Kids
bedtimehistorystories.com/tim-berners-lee-and-the-invention-of-the-internet-for-kids/
Listen to a podcast all about the invention of the internet here.

Timeline of Computer History
www.computerhistory.org/timeline/
Check out this timeline to learn more about the story of the computer.

INDEX